SPACE GUIDES

VOYAGE THROUGH SPACE

PETER GREGO

QED Publishing

Copyright © QED Publishing 2007

First published in the UK in 2007 by
QED Publishing
A Quarto Group company
226 City Road
London EC1V 2TT
www.qed-publishing.co.uk

A catalogue record for this book is available from the British Library.

ISBN 978-1-84538-686-3

Written by Peter Grego
Produced by Calcium
Editor Sarah Eason
Illustrations by Geoff Ward
Picture Researcher Maria Joannou

Publisher Steve Evans
Creative Director Zeta Davies
Senior Editor Hannah Ray

Printed and bound in China

Picture credits
Key: T = top, B = bottom, C = centre, L = left, R = right, FC = front cover, BC = back cover

Corbis/Bettmann 5, /Robert Landau 4–5, /Xinhua/Xinhua Photos 28; **ESA** 15T; **Getty
Images**/Photodisc FCB, 15B, 25T; **Lunar and Planetary Institute** /USRA 9T; **NASA** FCC,
1, 3, 6, 9B, 11B, 13B, 14, 16T, 16–17, 18, 19TL, 19B, 20, 21T, 21B, 23T, 23B, 24, 25B, 26,
26–27, 27, 29, /Charles M Duke Jr 22, /JPL 10–11, 11T, 16B, /JPL-Caltech 13T, /JPL/Cornell
University/Maas Digital BC, /Robert Markowitz/Mark Sowa 16TR, /NSSDC 7B, /USGS 8;
Peter Grego 4B; **Scaled Composites**/Jim Campbell/Aero-News Network 28–29; **Science
Photo Library**/Detlev Van Ravenswaay 12–13, /Mark Paternostro FCT.

Words in **bold** can be found in the Glossary on pages 30–31.

Contents

Flights of fancy

For thousands of years, people have dreamed of soaring up into the skies and travelling to the **Moon**, the **planets** and the stars.

The dream of space travel

Almost 2000 years ago, an Ancient Greek author called Lucian wrote about a journey to the Moon, Venus and the stars. In his story, 50 sailors crossing the Atlantic Ocean were carried into the skies in their ship by a mighty whirlwind. Lucian's description of the Earth seen from **space** – where cities, forests, mountains, rivers and seas were all clearly visible – is a lot like the description given by the first man to really travel into space, Yuri Gagarin, from his **spacecraft** in 1961.

⇧ The Moon rises over an Ancient Greek temple.

⇧ Lucian's travellers are whisked into space during a storm at sea.

From the Earth to the Moon is a famous book about space travel written by Jules Verne, in 1865. It describes how three people in a bullet-shaped craft are blasted into space by a gigantic cannon. This could never have worked – Verne's **astronauts** would have been flattened to the floor of their craft by the force of the blast-off.

In *The First Men in the Moon*, written by HG Wells in 1901, two people travel into space inside a sphere-shaped craft. It is steered by opening and closing shutters, operated from inside the craft. The explorers find the Moon a weird place, populated by strange, insect-like creatures called Selenites, who live beneath the surface in huge caves.

⬆ Passengers climb into Jules Verne's bullet-shaped space **capsule**.

Rocket power

In real life, rockets of all shapes and sizes have been sputtering across the skies for many hundreds of years. Ever since gunpowder was invented, more than 1000 years ago in China, it has been used to fuel rockets for war.

Liquid-fuelled rockets

In 1919, American scientist Robert Goddard worked out that it would be possible to launch a load high into the **atmosphere**, even into space. Many people laughed at the idea of using rockets to explore space. However, seven years later, Goddard launched the world's first liquid-fuelled rocket.

Robert Goddard launched the world's first liquid-fuelled rocket in 1926.

Rockets in action

All rockets – from the smallest fireworks to those powering the most advanced spacecraft – work in the same way. They burn fuel, which escapes as a fast stream of hot gases **from the rocket's base. The force, or** thrust**, of these gases pushes the rocket upwards.**

object flies in the opposite direction to the thrust

gases shoot out, producing thrust

The dawn of the Space Age

Rocket technology developed quickly in the mid-20th century. Once rockets became powerful enough to escape the Earth's **gravity**, scientists claimed they could carry **satellites,** or even humans, into **orbit**. Some people were not convinced. In 1956, one important **astronomer**, Richard Woolley, called the idea of space flight 'utter bilge'. He changed his mind one year later, when the first artificial satellite, Sputnik 1, glided above him in space. The **Space Age** had truly begun.

⇧ A toy balloon and a rocket fly in the same way: gases shoot out from the nozzle, producing thrust, and the object moves in the opposite direction to the thrust.

⇦ Sputnik 1, the first artificial satellite, was launched in 1957. It was little more than a radio inside a metal ball.

Close views of the Moon

The Moon, our nearest neighbour in space, was the first target for unmanned space **probes**. In September 1959, Russia's **Luna** 2 became the first probe to hit the surface of the Moon. As it hurtled down towards the surface, it sent pictures back to the Earth. A month later, Luna 3 took the first pictures of the far side of the Moon. No one had ever seen this side before, as it is always turned away from the Earth.

⇧ Thirty-five years after the first Luna 3 picture, this picture of the far side of the Moon was taken by the Clementine space probe. It shows the far side to be covered with **craters**, with fewer dark areas than the familiar near side.

More unmanned probes

During the 1960s, the United States investigated the Moon with three different unmanned **programmes** – Ranger, Lunar Orbiter and **Surveyor**. The Ranger probes took close-up pictures of the surface of the Moon before they crashed into it. The Lunar Orbiter probes photographed most of the Moon's surface from orbit. These images helped scientists to select the best landing sites for the Surveyor Moon probes, and for the later manned **Apollo** missions.

More recently, the United States has investigated the Moon with its Clementine and Lunar Prospector probes. The European Space Agency's SMART-1 probe completed a two-year-long survey of the Moon in 2006. More Moon probes are planned for the near future.

⇧ This image from Clementine shows the western half of the Moon. The colours match the heights of the landscape – blue areas are low ground, while red areas are high ground. The big, blue spot in the middle is the Eastern Sea, a giant crater more than three billion years old.

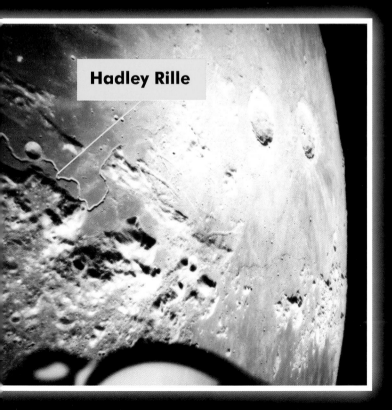

Hadley Rille

Ice at the Moon's poles?

Between January 1998 and July 1999, both the Clementine and Lunar Prospector space probes found signs of ice inside craters at the Moon's south pole. The floors of these craters are always in shadow, so they never get warmed by the Sun and any ice there never melts. Water will be a very valuable resource for future Moon explorers.

Amazing

Inner insights

The three planets nearest the Earth still have many secrets.

Mercury

Mercury is the least well-known of all the major planets. Its surface is full of craters, flat areas, ridges and cracks. **NASA**'s Messenger probe, launched in 2004, will fly past Mercury during 2008 and 2009. In 2011, it will begin a detailed, year-long survey of the planet.

⇦ Mercury, the planet nearest the Sun, was first photographed by NASA's **Mariner** 4 probe in 1975.

Venus

The surface of Venus, the second planet from the Sun, is hidden by its thick, cloudy atmosphere. In 1962, NASA's Mariner 2 became the first successful probe to reach Venus from Earth. The journey took three and a half months. It revealed that the planet's surface is incredibly hot. During the early 1990s, NASA's Magellan probe mapped Venus' surface.

Most of Venus is covered by smooth, hilly areas, from which rise three huge, mountainous **continents**. These are named Ishtar Terra, Lada Terra and Aphrodite Terra. The largest, Aphrodite Terra, stretches halfway around the planet. It is split by an enormous valley, called Diana Chasma. In places, the valley is 280km wide and 4km deep.

Mighty Maxwell

Rising above the continent Ishtar, Venus' Mount Maxwell is 11km high. This is much higher than Mount Everest, which at 8.85km is the tallest mountain on the Earth.

⬆ The Magellan probe took this picture of Venus' mighty Mount Maxwell.

⬇ NASA's Mars Odyssey is still studying the red planet closely.

Mars

It takes almost a year to reach Mars from Earth. Mariner 9 was the first probe to map Mars from orbit. It discovered that one side of the planet is heavily cratered, and that the other side is smoother, with some large volcanoes and one huge valley. Mars has been photographed by NASA's **Viking Orbiters**, Mars Global Surveyor and the space probe Mars Odyssey. A European Space Agency probe, Mars Express, has also studied the red planet in great detail.

Amazing asteroids

Between the planets Mars and Jupiter lies the **Asteroid** Belt. This broad zone contains tens of thousands of huge space rocks – material left over from when the Solar System was formed. Asteroids are tens of thousands of kilometres apart from each other.

Visiting probes

In the 1990s, NASA's Galileo space probe photographed three asteroids. The first was Gaspra, an angular mountain of rock about 18km long. The second was Ida, a potato-shaped asteroid that is 50km long. The third was Dactyl, a tiny asteroid in orbit around Ida. In 1997, NASA's NEAR-Shoemaker probe photographed an asteroid called Mathilde. In 2001, it arrived in orbit around an asteroid named Eros. It ended its mission by making a bumpy landing on Eros's rocky surface.

⇧ The icy nucleus of Halley's Comet was viewed by the Giotto space probe.

More missions

Japan's Hayabusa probe took detailed images of asteroid Itokawa in 2005. In 2010, it will bring back samples of dust collected from its surface. NASA's Dawn Mission will study asteroid Vesta between 2011 and 2015.

⇧ The NEAR-Shoemaker probe eventually landed on asteroid Eros. Eros is 30km long, and curvy in shape.

⇩ NASA's Stardust space probe took this image of Comet Wild 2 in 2004.

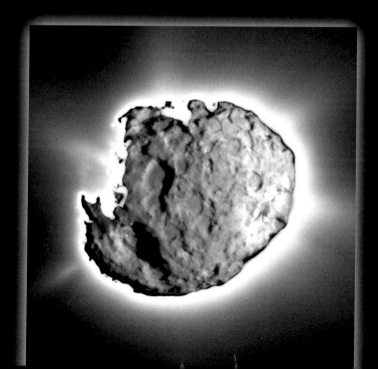

Cool comets

Comets are huge lumps of ice and rock. As they near the Sun, their surface heats up and the ice turns to gas, freeing the dust that was trapped inside. The centre of the comet, known as a nucleus, becomes surrounded by a hazy cloud called a coma, and a long tail of gas and dust streams away from it.

Halley's Comet

Halley's Comet has been seen and recorded by people on the Earth for more than 2000 years. It made its last passage through the Solar System in 1985–86. A European probe called Giotto sped into Halley's coma, and photographed its nucleus at close range.

Several other comets have been visited by probes. In 2004, NASA's Stardust probe collected material from the coma of Comet Wild 2, and returned it to the Earth to be studied.

Outward bound

Sending space probes beyond the Asteroid Belt and into the far reaches of the Solar System is a difficult task. They must be launched at the right time, at just the right speed, and in exactly the right direction so that they can eventually meet a planet hundreds of millions of kilometres away, at a time years in the future!

Key Concept

Probing space

Space probes use their engines to make small adjustments to their speed and direction. To make big changes, they use the gravity of other planets. Complicated mathematics and powerful computers are used by mission controllers on Earth to make the probes' movements as precise as possible.

On Jupiter's surface is ⇨ a large red oval called the Great Red Spot. It is a storm that has been raging in Jupiter's atmosphere for centuries.

Great Red Spot

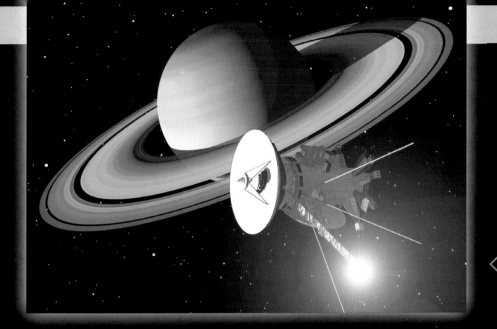

← NASA's Cassini space probe investigated Saturn, its rings and many moons.

Pioneer

Pioneer 10 was the first spacecraft to cross the Asteroid Belt. It encountered the giant planet, Jupiter, in December 1973. Pioneer 11 reached Jupiter a year later. It went on to fly past the ringed planet, Saturn, in September 1979.

Voyager

Voyager 1 has been travelling in space since it was launched in September 1977, making it the longest-lived space probe. It photographed the planet Jupiter in 1979 and the planet Saturn in 1980. Voyager 2 soon followed, and it took the first (and so far only) close-up photographs of Uranus in 1986 and Neptune in 1989.

⬇ Voyager's images of the outer planets and their satellites are amazingly sharp and colourful. The most distant planet, big blue Neptune, was imaged by Voyager 2 in 1989.

The images sent back by the Voyager probe show the clouds in the atmospheres of the outer planets as dark belts, bright zones, spots, ovals and swirls. They discovered hot, active volcanoes on Jupiter's innermost moon, Io, and cold fountains of **liquid nitrogen** on Neptune's moon, Triton. The probes also showed us craters on most of the moons and other features such as mountains and valleys.

Softlanders and rovers

An orbiting probe can only reveal a certain amount about a planet. A probe on the surface can discover a lot more about that planet's conditions, its climate and its rocks.

The Moon

A soft-landing probe is one that makes a controlled landing, rather than orbiting or crashing into a planet. Many American and Russian probes soft-landed on the Moon in the 1960s and 1970s. Russia's Luna probes sampled its surface, returned rocks and soil to the Earth and even travelled around the Moon by remote control.

Venus

The surface of Venus is hot enough to melt lead, and occasional volcanic eruptions result in planet-wide acid rainfall. The first softlander to successfully brave these terrible conditions was Russia's **Venera** 7, which parachuted down in 1970. It survived for just 23 minutes before being fried. Seven more Venera probes followed. Some returned pictures of the wild, volcanic landscape and sampled the soil. All were destroyed by the heat on the planet's surface.

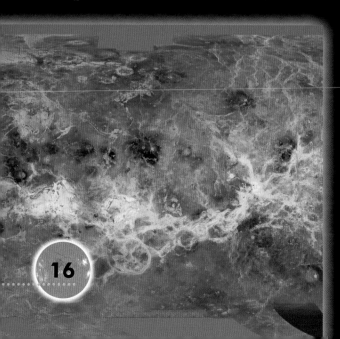

⇦ The Magellan space probe photographed Venus's surface in 1989.

Mars

In July 1976, NASA's Viking 1 lander touched down on Mars. Its photographs showed a landscape littered with dark rocks, and boulders embedded in dunes of orange dust. Two months later, Viking 2 landed. Both Vikings carried out experiments testing for life on the planet. The results proved nothing either way.

Rovers on Mars

Mars was revisited in July 1997, when NASA's Pathfinder mission delivered a **rover** called Sojourner. Two larger Mars rovers then followed in 2004 – Spirit and Opportunity. They took thousands of spectacular pictures, and made some fascinating discoveries. For example, Mars was found to have been a wet world in the distant past – water once flowed across its surface, producing layers of rock and unique minerals.

NASA's Spirit rover visited this field of bumpy boulders on Mars.

Human space flight

Sending people into space is an enormous challenge. Rockets are full of highly explosive fuel, there is no air supply in space, and on re-entry into the Earth's atmosphere any spacecraft becomes super-heated. All of this makes it very dangerous for astronauts.

First steps into space

In the 1960s, Russia and the USA competed to be the first to send a person into space. In 1961, Russia won the first part of the '**Space Race**'. In a tiny spaceship called **Vostok** 1, Yuri Gagarin orbited the Earth once. Four more Vostok flights followed, including the first whole day in space, and the first woman in space.

⇧ Yuri Gagarin (1934–1968) was the first person to orbit the Earth, in April 1961.

Biography

Sergei Korolyov (1907–1966)

Korolyov was the mastermind behind Russia's rocket programme. He helped to develop early missiles **which flew along a curved path, entering space at the highest point. He later improved the design of these missiles to make them capable of launching satellites and humans into orbit. He designed the Vostok and** Soyuz **space vehicles and the N1 Moon rocket. This was a powerful spacecraft, more than 100m tall, designed to carry Russian astronauts to the Moon.**

John Glenn was the first American to go into orbit, in 1962. In 1998 he then became the oldest person ever to go into space, aged 77. He spent nine days aboard the **Space Shuttle**.

Mission Mercury

The USA's **Mercury** programme ran alongside the Russian Vostok programme. The first two Mercury flights were small, 15-minute hops into space, while the final four missions went into orbit around Earth. Some of the Mercury astronauts went on to travel in space again. For example, in 1961 Al Shepard became the first Mercury astronaut. He went on to land and walk on the Moon in 1971.

The final Mercury ⇨ spacecraft was launched in 1963.

"One giant leap"

In 1961, the President of the USA, John F Kennedy, boldly set his country the goal of landing people on the Moon before 1970. To achieve this, the **Gemini** programme was set up.

Gemini

Gemini was designed to find answers to questions that would have to be solved in order to make a Moon landing possible. For example, how to **dock** spacecraft, alter their speed and orbit, and walk in space. The Gemini spacecraft flew 10 missions. On the second, astronaut Ed White made America's first space walk. For 22 minutes he floated way above the Pacific Ocean.

Ed White steps outside his Gemini craft to become the first American to walk in space.

Off to the Moon

Now the focus was on reaching the Moon itself. Apollo was the name of the USA's programme to land humans on the Moon. Saturn V was the giant rocket which carried astronauts to the Moon. It had three stages. The first stage heaved Apollo high above the Earth's atmosphere. The second stage pushed it into orbit, and the third stage launched it towards the Moon.

The huge Saturn V rocket lifts off from Cape Canaveral in the USA. Saturn V was a skyscraper of a rocket, 110m tall and 10m wide. ⇨

Neil Armstrong left the first human boot print in the Moon's soil. ⇧

Apollo 11

In July 1969, Apollo 11 carried astronauts Neil Armstrong, Buzz Aldrin and Michael Collins to the Moon. On 20 July, the landing craft Eagle touched down. Six hours later, Armstrong opened the hatch, climbed down the ladder and, placing his foot on the Moon's surface, said, "That's one small step for a man – one giant leap for mankind." Shortly afterwards, Aldrin stepped on to the Moon. These first **lunar** walks lasted two and a half hours. Armstrong and Aldrin set up some experiments and collected soil and rock. Collins remained in charge of the **command module**, Columbia, orbiting the Moon.

Lunar adventures

Following the success of Apollo 11, more Apollo missions came in quick succession.

Apollo 12, Apollo 13 and Apollo 14

In November 1969, astronauts Pete Conrad and Alan Bean of Apollo 12 landed on the Moon and found the Surveyor 3 probe that had landed two years earlier. They removed some of its parts so they could be examined back on Earth.

In 1970, an explosion on board Apollo 13 cancelled that mission, but later that year Apollo 14 landed safely. Television viewers watched as Al Shepard whacked a golf ball with a soil sampling stick.

⬆ Apollo 16 astronaut John Young leaped 1m high while saluting the USA's flag.

Apollo 15 astronauts drove to the edge of Hadley Rille, a magnificent valley.

Apollo 15, Apollo 16 and Apollo 17

Apollo's final three missions used an electric Moon buggy to explore further from their landing sites. In 1971, Dave Scott and Jim Irwin of Apollo 15 showed that objects of different weights fall at the same speed on the airless Moon. They dropped a hammer and a falcon's feather from the same height, and both hit the ground at the same time.

In 1972, John Young and Charlie Duke of Apollo 16 explored a crater whose inside walls were layered because of a build-up of lava flows.

Apollo 17 was the last mission to the Moon. Gene Cernan and Harrison Schmitt discovered some bright orange soil, which was later found to be made of tiny, coloured beads formed in the intense heat of a **meteorite** impact.

Key Concept

Bouncing on the Moon

Why do astronauts look as if they are moving in slow motion on the Moon? The force of gravity on the Moon is only one-sixth of its force on the Earth, so everything falls to the ground more slowly there.

Gene Cernan drove the lunar rover during Apollo 17.

Skylab and the Space Shuttle

After the Apollo missions, America's next space venture focused on a huge space station called **Skylab**.

Skylab

Skylab was a comfortable living and working space for three astronauts. An **observatory** was attached to its upper end, for studying the Sun. Electrical power was provided by **solar panels**. During Skylab's launch, in May 1973, one of its solar panels was torn away. The crew had to repair the damage.

Two more crews visited Skylab after the first crew. The last crew spent three months there, in 1974. NASA had planned to lift Skylab into a higher orbit using its new Space Shuttle. This was never possible, because the Shuttle programme was delayed. Skylab crashed to the Earth in 1979.

Skylab was ⇨ the USA's first space station.

An amazing machine

The Space Shuttle is the most complicated machine that humans have ever built. It is made up of more than 2.5 million parts, contains 370km of electrical wiring, and has 27 000 heat-resistant tiles covering its underside.

fuel tank

booster rockets

orbiter

rocket engine

⬆ The first Space Shuttle, Columbia, blasted off into space in April 1981.

Space Shuttle

NASA's Space Shuttle is a reusable vehicle lifted into orbit by its own rocket engines and two large rocket **boosters**. Once empty, the boosters fall back to Earth. They are recovered from the ocean and reused. The shuttle orbiter is a vehicle as big as a medium-sized airliner. It can stay in orbit for several weeks, and can carry up to seven astronauts.

Five Space Shuttles were built. Their large cargo bays have often been used to carry a science laboratory called Spacelab, which is used for experiments in space. Many satellites and space probes have also been launched from the cargo bay of the Shuttle, including the Hubble Space Telescope and the Galileo Jupiter probe. Space Shuttles are now being used to take materials and supplies to the ever-growing International Space Station (ISS).

⬅ As it returns to Earth, the Space Shuttle glides down through the air and lands on a runway.

United States

25

The International Space Station

The International Space Station (ISS) is a large base in which research about space takes place. It orbits Earth and is still being built. It is a joint project between five space agencies including NASA, the Canadian Space Agency, the Russian Federal Space Agency, the Japan Aerospace Exploration Agency and the European Space Agency.

⬇ The Space Shuttle Discovery prepares to dock as it approaches the International Space Station.

The ISS has been ⬆ growing steadily since 1998, as new **modules**, or parts, have been added. This is how it looked in 2005.

This shows how the International Space Station will look once it is completed.

An ambitious project

Since November 2000, at least two astronauts have been aboard the ISS at any time. It has been kept supplied with food, air, new crew and equipment by manned Soyuz spacecraft and the robotic Progress spacecraft (both from Russia), and by the Space Shuttle. The ISS will be completed by 2010.

When it is completed, the ISS will cover an area the size of a football field. Weighing about 450 tonnes – twice the weight of the Statue of Liberty – the ISS will be the largest structure ever to orbit the Earth. Several more modules will be added to it in the next few years. They will give crews of up to seven astronauts as much space to live and work in as the inside of a jumbo jet.

Project

Spot the ISS

The ISS orbits the Earth at a height of about 360km. Each orbit takes just over 90 minutes. This means that, in the right conditions, it can sometimes be seen in the evening sky as a bright, star-like object. Even with binoculars, you will not see more than a bright point of light, because it is so far away. You can find out when the ISS will be visible over your location, and exactly where to look for it, by visiting the website www.heavens-above.com.

Space for everyone

The USA has not been alone in developing its space programme. Other countries are also involved.

China

After many years of experience in launching rockets and satellites, China first launched an astronaut into space in October 2003. The **Shenzhou** space vehicle made 14 orbits around the Earth in less than a day. Two more, single-person space flights followed. China plans more complex missions in the future.

Russia

After abandoning its Moon landing programme in the late 1960s, Russia concentrated on a variety of Earth-orbiting space stations. The first, **Salyut** 1, was launched in 1971. Salyut 5 was launched in 1976. It successfully hosted two crews before crashing back down to Earth, in 1977.

A month later, Salyut 6 was sent into orbit. It had two docking ports – one for the Soyuz vehicle and another for a cargo vehicle to refuel it. Crews sometimes spent many months on board. In April 1982, Salyut 7 replaced Salyut 6. Over the next four years it was visited by 10 crews. It was eventually destroyed in 1991. Russia now plans to build a Moon base and send astronauts to Mars before 2030.

⇧ This is the capsule from China's Shenzhou spacecraft after it landed safely back on Chinese soil.

Mir

In February 1986, the core of a Russian space station called Mir was launched. Six different modules were connected to Mir during its lifetime in orbit, making it an impressively large object. It finally re-entered the Earth's atmosphere in March 2001.

This is the Russian space station, Mir. More than 50 visits were made to Mir, by several international crews, including six visits by the Space Shuttle.

Private space adventures

In September 2004, little **SpaceShipOne** lifted off above California to touch the edge of space. This was the first space mission paid for by an American citizen, rather than the US government. Its pilot was Mike Melvill. The company that built it is working on **SpaceShipTwo**, which will carry six fare-paying passengers and two pilots into space. Tickets will cost tens of thousands of dollars.

SpaceShipOne was the world's first private space plane.

Glossary

Apollo the USA's manned Moon landing programme of the 1960s and 1970s

asteroid a lump of rock in space

astronaut someone who has travelled more than 100km above the Earth

astronomer a scientist who studies space

atmosphere the layer of gases surrounding the Earth or another planet

booster the first, lift-off stage of a space rocket, or an extra rocket strapped to the side of the main rocket to provide extra thrust

capsule a small spacecraft

comet a city-sized ball of pieces of ice and rock in space. When a comet gets near the Sun it heats up and produces a long tail of gas and dust

command module a spacecraft that serves as a central base for astronauts

continent a large land mass

crater a circular, bowl-shaped hole in a planet's surface. Craters are blasted out by the impact of an asteroid, or the explosion of a volcano

dock the joining together of two spacecraft in space

gas a chemical which is not liquid or solid

Gemini a series of Earth-orbiting missions flown by the USA in the mid-1960s

gravity a force that acts throughout the Universe. The Earth's gravity holds everything to its surface, and the Sun's gravity holds the Earth in its orbit. The bigger the object, the more gravity it has

liquid nitrogen nitrogen gas that has cooled down so much it has become a fluid

Luna a series of 24 Russian Moon probes launched between 1959 and 1976

lunar an adjective to describe anything to do with the Moon

Mariner a series of 10 probes launched by NASA between 1962 and 1975

Mercury the USA's first space missions with a human crew

meteorite a small rock from outer space that has landed on the Earth

missile a rocket-powered weapon

module a special pod or enclosed compartment that is part of a space station or docked with it as a separate spacecraft

Moon   the Earth's only natural satellite. Other natural satellites are also known as moons (with a small 'm')

NASA   the letters stand for the National Aeronautics and Space Administration – the USA's space agency

observatory   where something is studied and scientific information is collected

orbit   the curved path taken by a satellite or spacecraft around a planet, or by a planet around the Sun

Pioneer   a series of 11 probes launched by NASA between 1958 and 1978

planet   a large, round object orbiting a star

probe   a spacecraft which gathers scientific information

programme   a planned mission to, or in, space

rover   an electric-powered wheeled vehicle used to explore other planets and the Moon

Salyut   Russia's space station programme

satellite   a natural or man-made object in orbit around a larger object. Natural satellites are also known as moons

Shenzhou   China's human space flight programme

Skylab   a large, 75-tonne space station occupied by three American crews between 1973 and 1974

solar panel   a device that converts sunlight into electricity

Soyuz   a series of crewed Russian spacecraft first launched in 1967

space   everything beyond the Earth's atmosphere

spacecraft   any vehicle that travels in space

SpaceShipOne   the first private spacecraft to enter space

SpaceShipTwo   an eight-seater spacecraft being developed for private space trips

Space Age   the era in which we now live, which began in October 1957 with the launch of Sputnik 1, the first artificial, or man-made, satellite

Space Race   the rivalry between the USA and Russia in all areas of space flight and exploration

Space Shuttle   a large, reusable winged spacecraft capable of carrying eight crew members into the Earth's orbit

Surveyor   a series of probes that softlanded on the Moon between 1966 and 1968

thrust   the force of hot gases produced when a rocket burns its fuel

Venera   a series of 16 Soviet probes to Venus launched between 1961 and 1983

Viking Orbiter   NASA's first robotic Mars missions

Vostok   Russia's first manned space programme

Voyager   two NASA space probes that travelled to the planets of the outer Solar System between 1979 and 1987

Index